# MAKE it WORK!
# SOUND

## Wendy Baker & Andrew Haslam

*written by*
Alexandra Parsons

*Photography*: Jon Barnes
*Series Consultant*: John Chaldecott
*Science Consultant*: Graham Peacock
Lecturer in Science Education at
Sheffield City Polytechnic

Scholastic Canada Ltd
123 Newkirk Road, Richmond Hill (Ontario) Canada

# MAKE it WORK!
## Other titles

Earth
Electricity
Plants

Published in Canada in 1993 by
Scholastic Canada Ltd.,
123 Newkirk Road, Richmond Hill, Ontario, Canada L4C 3G5

First published in Great Britain in 1992 by
Two-Can Publishing Ltd
346 Old Street
London EC1V 9NQ
in association with Scholastic Publications Ltd

Printed and bound in Hong Kong

2 4 6 8 10 9 7 5 3 1

**Canadian Cataloguing in Publication Data**

Baker, Wendy.
    Sound

(Make it work!)
ISBN  0-590-74522-0

1. Sound - Juvenile literature.
2. Sound - Experiments - Juvenile literature.
I. Haslam, Andrew.  II. Watts Claire.  III. Parsons, Alexandra.
IV. Barnes, Jon.  V. Title.  VI. Series: Baker, Wendy. Make it work!

QC225.5.B3 1993    j534'.078        C93-093608-6

Editor: Mike Hirst
Illustrator: Michael Ogden
Additional design: Belinda Webster

Thanks also to: Albert Baker, Catherine Bee,
Tony Ellis and everyone at Plough Studios.

# Contents

Words marked in **bold** in the text are
explained in the glossary.

# Be a Sound Scientist!

Did you know that sounds can blow out candles and crack glass? Or that bats use sounds to 'see' and find their way around? Have you ever wondered how different musical instruments produce completely different sounds?

## MAKE it WORK!

Start experimenting with sound. Find out how sound waves are made, how they travel, and some of the surprising things they can do.

## You will need

Most of the activities in this book use simple equipment, such as cardboard, glue and odds and ends. However, you will find some specialist equipment useful.

small hacksaw

bradawl

Investigating sound is part of the science of **physics**. Sound is a kind of energy, and it flows through the air in invisible waves. Everything that makes the air move, from the rustling of a leaf to the pounding of a big bass drum, sets off these **sound waves**, and creates a different kind of sound.

**Small hacksaw and bradawl** Some of the activities include making simple wooden supports or structures. You should always ask an adult to help if you are using saws, craft knives or other sharp tools.

**Cardboard tubes** These are handy for making musical instruments. You can use the insides of aluminum foil rolls or cardboard tubes made for carrying posters.

**Music manuscript paper** This is useful for writing down musical notes.

**Copper wire** This will be needed for making a simple radio and a telephone. Sound waves can be turned into an electrical pulse and then transferred along the wire. You can buy copper wire at any model shop or electrical supplier.

**Balloons and rubber bands** Some of the projects involve making a drum. Rubber from a balloon makes an excellent drumskin, and elastic bands will hold the rubber in place without tearing it.

tape recorder

SONY

CASSETTE-CORDER TCM-818

REC    PLAY    REW    FF    STOP/EJECT    PAUSE

earphones

**Tape recorder** A simple tape recorder is the most useful piece of equipment for sound experiments. Ideally, you will need a small tape recorder or Walkman, along with a microphone, small headphones and some blank cassettes.

**Plastic tubing and corks** These can be bought from shops that sell do-it-yourself wine-making equipment.

*When we hear a noise, the sound waves usually travel to our ears through air, but sound can move through other substances too. Whales, for instance, hear a wide range of sound waves that move through water.*

microphone

# 6 Sound Waves

The sounds we hear are caused when a moving object makes the air **vibrate**. These vibrations travel through the air in the form of waves, and are picked up by our ears as sounds. The shape of the sound wave depends on the **pitch** of the sound. Low-pitched sounds are deep and rumbling, like a big bass drum. High-pitched sounds are shrill and piercing, like a tin whistle.

## MAKE it WORK!

If you could see air, it would look like a great floating soup of gas particles. Low noises would make ripples a long way apart, and high-pitched sounds would make waves very close together. In fact, sound waves in air are invisible, but you can certainly prove they exist. Here are two ways to observe their effects.

### For the sugar drum you will need

| | |
|---|---|
| a cake tin | sugar |
| a wooden spoon | a balloon |
| large rubber bands | a baking tray |

**1** Cut out a circle of balloon rubber. Stretch it over the cake tin and fix it on with rubber bands.

**2** Sprinkle a little sugar on the top of the drum.

**3** Hold the baking tray above the drum and hit it with the wooden spoon. As the sound waves reach your ear, you hear the sound of the spoon on the tray. When those same waves hit the drumskin, they make it vibrate and you can see the sugar dancing up and down.

◀ Try holding the baking tray closer to the drum and then further away from it. Do you affect how much the sugar moves?

## For the sound cannon you will need

| | |
|---|---|
| a piece of plywood | candles |
| a plastic bag or a balloon | long nails |
| a cardboard tube | rubber bands |

**1** Ask an adult to help you hammer three nails through a piece of wood. Turn the wood over and push a candle onto each nail.

**2** Take a length of cardboard tube. The inside of a roll of aluminum foil will do fine.

**3** Stretch a circle of balloon rubber or plastic bag over each end of the tube and secure them with rubber bands.

**4** Make a little hole in the plastic stretched over one end of the tube.

**5** Ask an adult to light the candles.

**6** Point the end of your sound cannon with the hole in it at one of the candles. Hold it just a short distance away.

**7** Tap the other end with your finger. The vibrations you set up by tapping the drumskin travel to your ear as sound waves. The same vibrations move down the tube and push the air through the little hole at the opposite end, blowing out the candles.

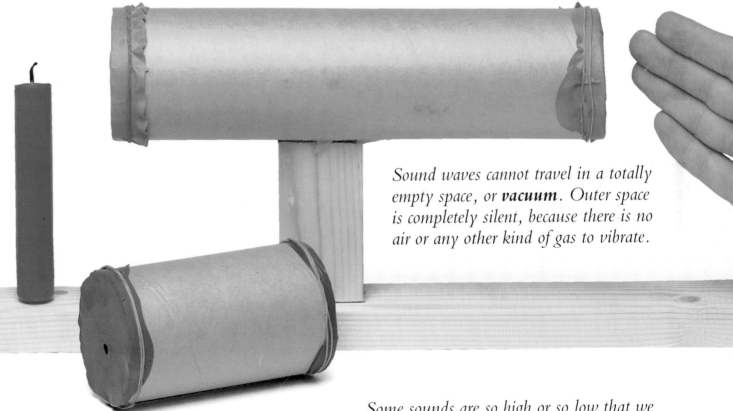

*Sound waves cannot travel in a totally empty space, or* **vacuum**. *Outer space is completely silent, because there is no air or any other kind of gas to vibrate.*

*Scientists measure sound waves by the number of times they make the air vibrate each second. One vibration, or cycle, per second is called one* **Hertz** *(Hz). When you hear a 700 Hz noise, the sound waves are hitting your eardrum 700 times per second. Scientists also measure other types of wave, such as light waves or radio waves, in Hertz.*

*Some sounds are so high or so low that we cannot hear them. We are unable to make out sounds that have a frequency above roughly 20,000 Hz, or below 20 Hz. However, many animals have a much wider hearing range than humans. Bats and dogs, for instance, both pick up much higher sounds than we do. There are even special dog whistles that give out a high-pitched noise which only dogs can hear.*

Our ears are specially designed to pick up vibrations in the air and turn them into **nerve pulses** which our brains understand as sounds. The working parts of our ears are actually inside our skulls. The whorled flaps that stick out on either side of the head are just funnels to collect sounds and pass them along to the **ear-drum**.

### MAKE it WORK!

Make your own working model of a human ear. The ear-drum vibrates, moving three small bones inside the middle ear. These bones then move a fluid in a curly pipe called the **cochlea**. The cochlea is lined with minute hairs, and as the fluid moves, so do the hairs, sending tiny sound impulses along the nerves to the brain.

### You will need

| | |
|---|---|
| wood and nails | corks |
| thin dowels | thin card |
| hairy string | stiff wire |
| plastic tubing | a balloon |
| paper fasteners | tape and glue |
| a cardboard tube | food colouring |

**1** Take a length of cardboard tube and cut away part of one side as shown. Place a circle of balloon rubber over one end, fixing it securely with rubber bands. This will be the ear-drum.

**2** Make the bones that are connected to the ear-drum – the **malleus**, the **incus** and the **stapes**. In the model, these bones are made from pieces of card. Cut out and fold the shapes shown in the photograph. Then glue and fix them together with paper fasteners as shown in the diagram on the right.

**3** Stick the flaps of the cardboard malleus onto the rubber of the ear-drum with rubber solution glue.

ear-drum

**4** Cut a length of bendy, transparent plastic tubing and thread a slightly shorter length of string through it. Use a rough kind of twine, made out of natural fibres. The plastic tube is the cochlea, and the fibres on the twine are the tiny hairs that send nerve pulses to the brain.

Making the middle ear

**6** Ask an adult to drill a narrow hole in one cork stopper and push a small piece of thin copper tubing through the hole. Then thread the stiff wire through the copper tubing and stick it into the thinner piece of cork as shown.

**7** Push the cork and wire mechanism into one end of the plastic tube. Almost fill the tube with water, and use the remaining cork to stop up the other end.

**5** Connect the bones to the cochlea. Whittle down a piece of cork until it fits exactly inside the plastic tube. Cut two other corks so that they will stop up the ends of the tube.

**8** Make a wooden base and put the separate parts of the model together, holding them up on dowelling supports. Glue the wire to the cardboard stapes and twist the cochlea into a spiral shape.

**9** Make an ear shape from coloured card, and glue it to the free end of the tube.

malleus

organs of balance

stapes

incus

cochlea

▲ Ears are important for other things besides hearing. They also help us to keep our balance. You can make your model ear even more authentic by adding semicircular tubes filled with coloured water. These represent the fluid canals that are our organs of balance.

**Operating the model**
To watch your model work, tap lightly on the inside of the ear-drum, to make it vibrate as though it had been hit by a sound wave. Can you see what happens to the little hairs on the string inside the cochlea tube?

Have you ever wondered why you have two ears, placed on either side of your head? It's so that you can tell which direction a sound is coming from. Unless a sound is directly in front of you, one ear will always pick it up slightly before the other. Your brain uses the different information coming from each ear to work out the direction of the sound.

**MAKE it WORK!**

To find out where a sound is coming from, we usually turn our heads until both ears hear the sound equally and the sound is 'in focus'. Make this game and test whether your ears have a good sense of direction.

**You will need**
large sheets of
    coloured paper or thin card
a pencil, scissors and string
some friends
a blindfold

**1** Take a large sheet of paper and fold it in half. Attach a pencil to one end of a length of string and fasten the other end of the string to the piece of paper at a corner along the fold.

**2** Use the string and pencil like a compass to draw an arc. Then cut along the line, and unfold the paper to give you a semicircle. Repeat with another piece of paper the same size.

**3** Put the two semicircles together to make a whole circle. If you wish, you can divide the semicircles into segments and make alternate colours. Then cut out a paper arrow.

*Noises behind us sound muffled because our ears are curved forward to receive sounds from in front.*

Compared with light, sound waves travel quite slowly – light moves almost a million times faster than sound. That's why, during a thunder storm, we see a flash of lightning before we hear a clap of thunder, even though they are both, in fact, caused at exactly the same time.

### ▼ Playing the game

This is a game for three to five people. One person volunteers to wear a blindfold and sits at the centre of the circle. The other players stand or sit silently around the circle, and someone makes a gentle noise, such as clicking their fingers. The blindfolded person points the arrow in the direction they think the sound is coming from. Each player can have a go at being blindfolded. Whose ears have the best sense of direction?

▶ Ask a friend to stand at the other side of a field or playground, holding up a handkerchief. (The further away your friend, the better the experiment will work – a pair of binoculars could help here.) Tell the friend to shout and drop the handkerchief at exactly the same time. You should see the handkerchief begin to fall before you hear the shout.

Instead of card, you could make the circle out of felt or other material.

*Children have a wider range of hearing than adults. They can hear higher-pitched sounds.*

# 12 Reflecting Sound

Our ears pick up the sound waves travelling through air, but, in fact, sound waves can travel through all kinds of substances. Whales, for instance, hear sound waves in water. And sound can even pass through a solid object, such as wood, if it is made to vibrate. However, sound doesn't pass very well from one type of substance to another. So if you shout at a brick wall, not much of the sound gets from the air through the wall and out at the other side. Instead, the sound waves bounce back off the wall again.

## Angles of bounce

Sound waves in air will bounce off a flat, solid object at the same angle as they hit it – just like a ball bouncing off a wall. If, instead of a flat, solid object, the sound waves are bounced onto a surface that is soft or bumpy, the waves will break up and fade away.

## You will need

thick card
a craft knife
cardboard tubes
a tape recorder and microphone
a clock or watch with a very soft tick

an egg tray
Plasticine

## MAKE it WORK!

In this experiment, you can control the path of the sound waves by directing them along cardboard tubes. The tubes hold the sound together, making it louder because the sound waves can't spread out and get lost in the air around them.

**1** Take four equal lengths of cardboard tube, and cut three squares out of a piece of thick, smooth (not corrugated) card.

**2** Use Plasticine to fix the cardboard tubes and the squares of card in position as shown below. Each tube must be placed at exactly the same angle to the cards.

**3** Measure the distance in a straight line between one end of the zig-zag and the other.

**4** Set up your clock or watch away from the tubes, and record it ticking across the distance you have measured. Your microphone will pick up only a faint sound, or no sound at all.

**5** Now position the clock at one end of the zig-zag, and record the sound that comes out of the other end.

**6** If all of your tubes are positioned at the same angle, you should be able to record the ticking sound clearly. The sound waves travel down one tube, bounce on and off the reflector card at the end, and carry on back down the next tube.

**7** Try altering the position of the tubes, and record what happens. If the angles don't match, the sound waves just spread out into the surrounding air, getting weaker and weaker.

### Deadening sound
Put the cardboard tubes back in their original positions and then experiment with reflector cards made out of different materials. Cut some squares out of an old egg carton, so you can test the effect that a bumpy, uneven surface has on sound waves.

*Architects use their knowledge of bouncing sound waves when they design new buildings. A noisy restaurant can be made much quieter by covering the floor, walls and ceiling with soft fabrics and bumpy surfaces to deaden the sound. But the stage and walls of a concert hall can be built to reflect sound waves, so that the music travels clearly towards where the audience is sitting.*

Before electronic hearing aids were invented, people who had difficulty hearing used an ear trumpet. They put the narrow end to their ear, and if someone spoke clearly into the wide end, their voice could be heard more clearly. The trumpet **amplified** the sounds, or made them louder.

The simplest amplifier is a big cone. It can be used to send out sounds or to listen to them. When it is used to send sounds, the cone holds the sound waves together, so they don't spread out in the air so quickly. When a cone is used to listen, like the old-fashioned ear trumpet, it collects sound waves from the air and directs them into the ear so they sound louder.

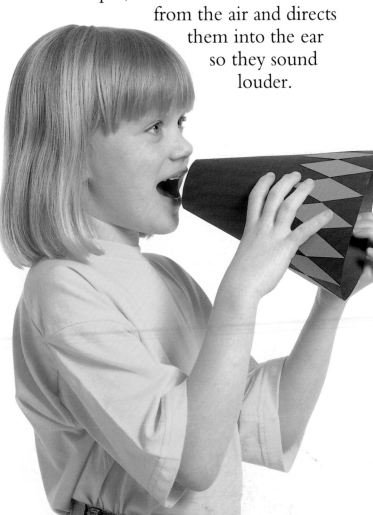

## You will need

thin coloured card
flexible plastic tubing
two plastic funnels

glue
scissors

## MAKE it WORK!

You can make some instruments to amplify sound waves. Doctors use a special listening device called a **stethoscope** to hear inside their patients' bodies. Normally, we can't hear the quiet gurgles and thumps made by our bodies, but the stethoscope picks up the sound waves and leads them directly into the doctor's ear.

### Making a stethoscope

Take a length of plastic tubing that fits neatly over the narrow ends of the two funnels. Attach a funnel to each end.

Ask a friend to hold one funnel over his or her chest, put the other funnel to your ear and listen carefully. You will be able to hear your friend's heartbeat.

*The loudness of a sound is measured by the force with which the sound wave pushes the air. The units of measurement are called* **decibels** *– named after the scientist and inventor Alexander Graham Bell.*

## Making an ear trumpet and a megaphone

**1** Take a sheet of thin coloured card and roll it into a cone shape. One end should be wide, to collect sound waves, and the other end should be narrow enough to fit into your ear.

**2** Make the megaphone cone in the same way, but the narrow end should be slightly larger so that you can speak into it.

**3** Once you have the right cone shapes, tape or glue the card. Trim the ends and decorate the cones with coloured paper shapes of bright, contrasting colours.

**4** To test the megaphone ask a friend to stand so far away that you can't hear each other's voices. Then speak normally through the cone.

**5** To test the ear trumpet, point it towards a quiet noise. You will be able to hear the sound more clearly. **Be careful**! Never shout down an ear trumpet at anyone. You could damage their ear-drum.

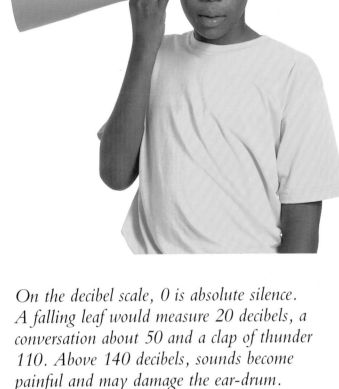

*On the decibel scale, 0 is absolute silence. A falling leaf would measure 20 decibels, a conversation about 50 and a clap of thunder 110. Above 140 decibels, sounds become painful and may damage the ear-drum.*

# 16 Travelling Sound

Sound waves can be converted from sound energy into electrical energy and back again. That's how sounds can be carried over long distances by the telephone. The sound waves are turned into an electrical pulse, which travels down the telephone wire to the receiver – just like the nerve pulses that carry sounds from our ears to our brains.

## MAKE it WORK!

Modern telephones use carbon granules to convert sound waves into electrical pulses. However, you can make a simple telephone for yourself, using magnets and copper wire.

When you speak into the paper cup, the paper drumskin vibrates, so that the magnet bounces up and down inside a coil of copper wire. This movement causes an electrical charge, which passes down the wire. At the other end of the wire, another copper coil, magnet and drumskin convert the electrical pulses back into sound waves

edging piece for cradle

drumskin

reel

coil

star shape for cradle

## You will need

| | |
|---|---|
| two large plastic or paper cups | scissors |
| airmail writing paper | coloured card |
| | |
| insulated electrical wire | glue and tape |
| single-strand copper wire | |
| two cylinder magnets (from a model shop) | |

**1** First make two copper coils. Cut and glue the pieces of card as shown to make reels that will fit neatly around the cylinder magnets. Wind a long piece of copper wire around each reel forty or fifty times.

**4** Cut a drumskin out of thin airmail paper. Snip the edges so that you can glue it to fit tautly inside the rim of each paper cup.

**5** Now assemble each paper-cup receiver. First, glue the reel and coil into the cradle. Fix the cradles inside the paper cups as shown.

**6** Glue the magnet to a small circle of card and glue this card in turn onto the back of the paper drumskin. Then glue the drumskin in place. Check that the magnet fits neatly inside the reel, but has room to move freely up and down.

**7** To talk into the telephone, cup your hands around the cup and talk right at the drumskin. To listen, hold the drumskin up to your ear.

*Some modern telephone wires now use optical fibres, which transfer sound not as an electrical pulse, but as a high-speed beam of light.*

▲ **cylinder magnet attached to drumskin**

▲ **cradle**

▶ If your connecting wire is long enough, you can use this phone to speak to a friend in another room.

**2** Cut star shapes and edging pieces out of thick card as shown. Then put them together to make cradles for the reels and copper coils.

**3** Pass two equal lengths of insulated wire through the bottom of each paper cup and connect them to the ends of the copper coils.

# 18  Recording Sound

Sound was first recorded by a machine called a **phonograph** – a kind of early record player. A needle was attached to a drumskin, stretched across the narrow end of a sound horn. When someone shouted into the horn, the needle would vibrate. As the needle bounced up and down, it recorded the sound waves as grooves on a cylinder coated in wax or tin foil.

## Tape recordings

Sound recording techniques have come a long way since the days of the phonograph. Sound is now recorded onto magnetic tape. In the tape recorder, the sound waves are turned into electric impulses. These are stored on the tape as a sequence of different magnetic blips.

## MAKE it WORK!

A microphone is a kind of electric ear. It turns sound waves into electric signals. However, it may have trouble picking up sounds not made close by. You can improve a simple microphone by putting it inside an umbrella! Its shape will collect the sound waves and reflect them back to the microphone.

To play a phonograph recording, the cylinder was rotated underneath the needle. The pattern of bumps and dips in the cylinder grooves vibrated the needle, which the drumskin and sound horn turned back into sound waves.

## You will need

| | |
|---|---|
| a tape recorder | tapes |
| a microphone | earphones |
| sticky tape | an umbrella |
| three or four friends who can sing | |

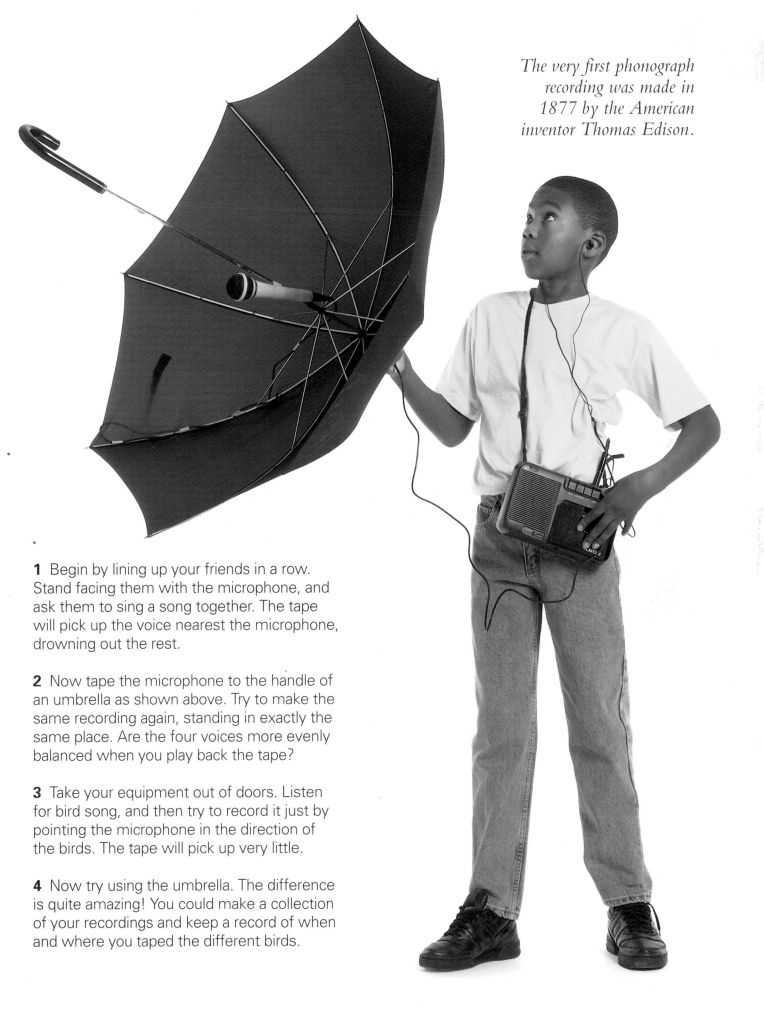

**1** Begin by lining up your friends in a row. Stand facing them with the microphone, and ask them to sing a song together. The tape will pick up the voice nearest the microphone, drowning out the rest.

**2** Now tape the microphone to the handle of an umbrella as shown above. Try to make the same recording again, standing in exactly the same place. Are the four voices more evenly balanced when you play back the tape?

**3** Take your equipment out of doors. Listen for bird song, and then try to record it just by pointing the microphone in the direction of the birds. The tape will pick up very little.

**4** Now try using the umbrella. The difference is quite amazing! You could make a collection of your recordings and keep a record of when and where you taped the different birds.

# 20 Ultrasound

Bats have poor eyesight, but a very good sense of hearing. They can hear ultrasound – very high-pitched sounds, way beyond the human hearing range. Bats can use these sounds to find their way around the dark caves where they live. When a bat makes a high-pitched squeak, the sound bounces off the wall of the cave and returns to the bat as an ultrasound echo. From the amount of time between the squeak and echo, the bat can tell how far away the wall is.

**1** Take a rectangle of black card, fold it over and draw the shape of a bat wing as shown above. Cut out this shape, unfold the card and you'll have a symmetrical bat. Make twelve bats, three for each player.

**2** Take a piece of cork for each bat. Stick a Velcro dot underneath the cork, and then glue the cork itself to the side of the bat, with the ears facing up.

### MAKE it WORK!
Make this ultrasound game, and turn over the page to find out how to play. The board is a dark cave, criss-crossed by sound waves. The playing pieces are bats, and they use the sound waves as flight paths to move from one wall of the cave to another.

### You will need
corks and glue
wooden dowels
a round toothpick
thick coloured card
thin coloured card
a craft knife or scissors
Velcro dots or paper clips

**3** If you can't get hold of any Velcro dots or corks, simply attach a paper clip to the bottom of each bat as shown on the left. Velcro bats will sit on top of the sound waves. Paper-clip bats, on the other hand, hang beneath the sound waves.

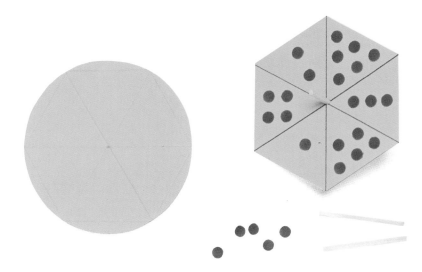

**4** Choose a colour for each player. Put different coloured dots on each player's bats.

**5** Make the spinner. Draw a circle and divide it into six segments. Snip off the sides to make a hexagon, and stick on the coloured dots.

**6** Poke a round toothpick through the middle of the spinner. Twirl it around and it will come to rest on one of the six sides.

### Assembling the board

**7** Now make the board. Cut a large square of card for the base, and glue a length of dowel standing on its end in each corner.

**8** Cut four strips of thick card, each as long as a side of the baseboard. Divide each strip into sixteen small squares. Glue these strips in place on the wooden dowels, to make the raised section of the board as shown in the diagram above.

**9** At each corner of the raised section put a square of paper in one of the players' colours.

**10** Cut six long strips of thin card to represent sound waves. If you have made Velcro bats, mark off the sound waves with Velcro spots. If you have paper-clip bats, stick a row of coloured-paper spots on each sound wave.

**11** Put the sound waves in position, stretching the strips of thin card in wavy lines across the board from one raised side to the other. In some places, you will have to add extra dowel supports to hold up the sound waves and help them to keep their shape.

## Playing the ultrasound game

In this game, each player has three bats which set off from the corner marked in their colour. The aim of the game is to make as many flights as possible from one side of the board to the other within a set time limit.

**1** Decide on your time limit for the game – say fifteen or thirty minutes.

**2** Spin the spinner. The player with the highest number starts, and play passes to the left.

**3** Players spin in turn and move their bats along the edge of the playing board. When a bat reaches the beginning of a sound wave, it may cross the cave, spot by spot.

**4** Each time a bat completes a crossing, the player puts a scoring counter or old button in his or her corner of the cave. The bat continues around the edge of the board until it reaches another sound wave.

**5** Players may use a spin to move any one of their bats. They may also split the score and move more than one bat.

**6** When two bats going in opposite directions meet, they cannot pass one another. The two players both spin the spinner. The bat with the higher score continues its flight, but the lower-scoring bat falls off and goes back to its corner.

**7** At the end of the agreed time limit, the players have a race back to their corners, completing any waves they have already started. The first player to bring all three bats home gets three extra counters.

**8** The winner is the player who has collected the most counters.

# 24 Radio Waves

The electronic pulses of the telephone wire aren't our only way of making sound travel over long distances. Sound waves can also be converted into **radio waves**. With a radio transmitter and receiver, these waves can be made to carry sound all around the world.

## MAKE it WORK!

Radio waves are rays of electromagnetic energy, which we cannot see or feel. They travel at the speed of light, through air, solid objects and even completely empty space. At a radio transmitting station, sounds are turned first into electrical impulses, and then sent out as radio waves. A radio receiver picks up these waves and turns them back into sounds again.

Make your own simple radio receiver and see which radio signals you can tune into.

### You will need

| | |
|---|---|
| corks | steel wire |
| a bradawl | a paper clip |
| wood glue | rubber bands |
| balsa wood | headphones |
| a cardboard tube | a sharp craft knife |

nuts, bolts and washers
a diode, from a model shop
two lengths of single-strand electrical wire
bare copper wire (without insulating laquer)

**1** Ask an adult to help you cut a square piece of balsa wood for the base board.

**2** Cut two more pieces of wood to be the supports for the cardboard tube, as shown above. Glue them into position on the base.

**3** With the help of an adult, cut four slices of cork and glue them beneath the base as feet.

**4** Drill four holes along the front of the base, and screw bolts into three of them as shown. Put a washer between each bolt head and the base, and secure the bolts beneath the board with nuts.

**5** Put the plug of the earphones into the remaining hole.

**6** Wind the bare copper wire around the cardboard tube. Make the coil as tight as possible. The turns should touch one another, without actually overlapping.

▼ The **diode** directs the electromagnetic impulses picked up by the **aerial**, forcing them along the wire and into the headphones.

**9** Fix the diode between the middle bolt and unconnected outer bolt.

**10** Push the steel wire through the balsa wood supports in front of the copper coil, and bend it round to connect it to the diode as shown.

**11** Attach a length of insulated wire to each of the outer bolts. To make an aerial, take the wire connected to the diode, and fasten it to a long metal object, such as a tap or radiator.

▼ To operate the radio, slip a paper clip onto the steel wire, twist it forward, and move it along the copper coil. Grasp the unconnected end of insulated wire and put on the headphones.

**7** Secure the coil with rubber bands at both ends. Leave one short end of the coil free.

**8** Take the other, longer end of copper wire. Fix it beneath one of the outer bolts and around the earphone plug, attaching the very end of the wire to the middle bolt.

Every sound has its own pitch – high or low – depending on the shape of the sound wave. To a musician, each different pitch of sound makes an individual musical **note**.

Musical notes can be arranged in a special pattern called a **scale**. There are many kinds of scale, but the most common is called an **octave**. It is made up of eight notes that move higher and lower step by step. The notes on a piano are set out in octaves – and you might have sung the notes of an octave yourself: do, ray, me, fa, so, la, te, do.

## MAKE it WORK!

Make your own instruments that play the notes of the octave scale. You will need to learn the notes of the octave, or copy them from a musical instrument – an instrument with fixed notes, such as a piano or electric organ, is easiest. If you don't play an instrument yourself, you could ask for some help from a friend or adult who does.

You can experiment with both a metal twanger (below) and a bottle organ (right) to create instruments that play the notes of the octave scale.

## For the bottle organ you will need

food colouring          a long nail
a musical instrument
eight bottles of the same size and shape

**1** Collect eight identical bottles and set them up in a line. (Juice bottles are probably the easiest to use.) Hit each bottle with a long nail. They should all make the same sound.

**2** Take a jug of water and pour a little into the first bottle. If you are going to copy a scale from a musical instrument, play the first note of the scale. Then gently hit the top of the bottle with the nail to see if it sounds close to the note you have just played. Gradually adjust the water level in the bottle until both bottle and instrument give out exactly the same note.

**3** Follow the same method for each bottle until you have made a complete octave.

**4** When you are happy with the sound of the bottle organ, dye the water in each bottle with a few drops of food colouring. You will be able to see the different notes more easily.

**5** Now try to pick out a tune – something simple, such as 'Three Blind Mice'.

## Making a metal twanger

You can make another octave scale using strips of metal or thick steel wire, fixed to a solid block of wood beneath the heads of screws.

### For a brass twanger you will need

| | |
|---|---|
| a small block of wood | a file |
| a thin strip of brass | screws |
| a small hacksaw | a screwdriver |
| a bradawl | a hammer |

**Be careful!** To make a twanger, you will have to use sharp tools. Ask an adult to help you.

**1** Cut a short strip of brass and hold it in place on the block of wood. Pluck the end of the strip, and it will make a twanging noise.

**2** Now cut seven other strips of brass, each slightly longer than the last. Experiment with different lengths until you can arrange the strips on the block of wood to make an octave.

▲ You can make another simple twanger using thick steel wire, held in place by U-shaped tacks.

**3** File the sharp edges off the strips of brass and fix them in place. Mark where the screws should go with a bradawl and then twist them into the wood with the screwdriver.

▼ Musicians check the pitch of an instrument with a tuning fork. When the fork is struck, its prongs vibrate at a regular speed, sending out a clear, single note that never changes.

*A lot of the pop music and classical music we hear is based on the eight-note scale, but some styles use totally different scales. Much Chinese music is based on a sequence of five notes; Indian classical music often uses a twenty-two note scale.*

How many kinds of musical instrument can you think of? They come in many different shapes and sizes, from mouth organ to double bass, synthesizer to tom-tom drums. Yet they all have one thing in common – they create sounds by making the air vibrate.

### Musical pipes

Many musical instruments have pipes – organ pipes, the curly tube of a French horn or the thin, straight pipe of a tin whistle. Inside each pipe is a column of air. When this air is made to vibrate in a certain way, the pipe plays a note.

### MAKE it WORK!

Make your own music using pipe instruments. Panpipes are a set of pipes of different lengths, each of which makes a different note. Tubular bells are metal pipes that hang free and make a bell-like sound when hit with a large nail.

### To make the panpipes

**1** Cut eight straws or eight pieces of plastic tubing to different lengths. Tape them together, stepped at one end, as shown above.

**2** To play the panpipes, blow gently across the top of each pipe to produce a note. You will find that longer pipes make a lower note than shorter pipes.

### You will need
glue
wood
copper pipe
nylon fishing line
bendy plastic tubing          sticky tape
screws and a screwdriver      a hacksaw
plastic drinking straws       a large nail

**3** Experiment with the lengths of the pipes. Can you make octave scales like those on the previous page?

## To make the tubular bells

**1** Ask an adult to help you cut eight lengths of copper pipe. Each one should be a little longer than the one before, so that you can arrange them in a stepped pattern.

**4** Hang the copper pipes from the top piece of the frame using lengths of strong fishing twine. Thread the twine through the holes you have drilled, and attach it to the copper pipes with strong plastic tape.

*Pipes that give out a sound when they are struck are **percussion instruments**. Pipes that are blown to make music are **wind instruments**.*

**2** Make a wooden frame. It should be big enough to hold the copper pipes hanging at well-spaced intervals. You will need a flat base, two side pieces and a top piece.

**3** Drill eight holes at equal spaces through the top piece of the frame. Then put the frame together, cutting the side supports into the baseboard. Glue the joints and secure them firmly with screws.

**5** Arrange the pipes in order of length. Strike them with a metal nail to make them chime.

Because the pipes hang free, they carry on vibrating after they have been hit, and the sound lingers on. Try striking a note and then grasp hold of the pipe tightly with your hand. The sound dies straight away. As soon as you stop the pipe vibrating, the sound waves stop being produced too.

# 30 Slides & Air Holes

Unlike our panpipes, many instruments create sounds by vibrating the air in just a single pipe. In these instruments, the different notes are made by altering the length of the main pipe. Recorders have holes to let the air out in different places. A brass trombone has a slide that moves up and down to change its length.

**1** Take a length of copper pipe or a section of cardboard tubing. Leave a space at one end, and then use a bradawl to mark out evenly-spaced holes along the rest of the pipe.

**2** Ask an adult to drill the holes you have marked. At the end of the pipe nearest the mouthpiece, cut a blowhole, following the diagram on the right.

▼ tin whistle

▼ brass-pipe recorder

▼ cardboard-tube recorder

▲ ▼ slide whistles

### Make a recorder
A recorder player makes different notes by putting fingers over the holes to make the air travel further down the pipe.

### To make a recorder you will need
wooden dowelling            a cork
a tube of thick cardboard   copper pipe
a bradawl, drill and hacksaw

**3** Cut a slice off a piece of cork to give it a flat edge. Then sandpaper the cork to make it smooth and fit snugly inside the top of the recorder pipe. Position the flat side face up, opposite the blowhole and finger holes.

**4** Now blow gently into the mouthpiece. The cork and the blowhole are shaped so that the air inside the recorder is vibrated.

Making the recorder mouthpiece

### To make a slide whistle you will need
a copper pipe          sticky tape or glue
wooden dowelling
the mouthpiece from a party blower

**1** Ask an adult to cut a length of pipe.

**2** Cut a piece of dowelling about 15 cm (6 in) longer than the pipe. The dowelling should fit inside the pipe as tightly as possible, while still being able to move easily.

**3** Fit the party blower mouthpiece to one end, using glue or sticky tape. If you cannot find a mouthpiece that fits, you can still play the slide whistle by blowing across the top of the pipe.

### Make a slide whistle
Slide whistles make a weird wailing sound. The length of the pipe is changed by pushing a dowel in and out. There are no separate notes, just a continuous rising and falling tone.

*The width of a pipe, as well as its length, will affect the sound it makes. Narrow pipes make higher-pitched notes than wide pipes.*

Shakers and rattles are all percussion instruments. They are often used to stress the rhythm of a piece of music and give it a strong beat. Shakers, like our paper-cup shakers, make the air vibrate by moving loose objects inside an enclosed space. Rasps and rattles rub together two rough surfaces to set up sound vibrations.

**To make the shakers you will need**

paper cups              sticky tape
plastic bottles with lids
rice, lentils, chickpeas, beads and pebbles

**1** To make a paper-cup shaker, put a handful of rice or lentils into one cup. Turn another cup upside down and tape the two cups together, rim to rim.

**2** To make plastic-bottle shakers, simply pour a handful of beads or chickpeas into the bottle, and put the lid on tight. You could decorate the shakers with coloured paper if you wish.

**3** Try making shakers with different-sized bottles. You will find that larger bottles which hold more air make deeper sounds.

**MAKE it WORK!**
Make your own collection of different shakers and rattles. They could make a whole variety of sounds, depending on the amount of air that is made to vibrate and the kinds of surface that are rubbed up against one another.

▲ Experiment with different fillings for your shakers. You will find that paper-cup shakers with lentils make a softer sound than plastic bottles with chickpeas.

## To make a rattle you will need

a wooden dowel or stick          a hammer
several long, thin nails          enamel paints
metal bottle tops                a large, thick nail

**1** Punch a hole in the middle of each bottle top using the hammer and large nail. **Be careful** as you do this! Make sure that you've put a piece of old wood underneath the bottle top, and concentrate so that you don't hit your fingers.

**2** Slip four bottle tops onto each thin nail. (If you like, you could give them a coat of enamel paint first.) Then hammer the nails into the dowel or stick, making sure the tops can rattle freely.

## To make a sandpaper rasp you will need

two blocks of wood          thumb tacks
two sheets of sandpaper

**1** Pin the sandpaper to the blocks of wood as shown below.

**2** To play the rasp, rub the two sandpaper surfaces together. It makes a soft, grating noise.

## To make a wooden rasp you will need

a block of soft wood          a large nail
a small hacksaw

**1** Ask an adult to help you cut a zig-zag shape along the top of the block of wood as shown.

**2** To play the wooden rasp, run the nail back and forth along the uneven surface. It makes a harsher sound than the sandpaper rasp.

Both of these rasps make quite soft noises. Unlike the shakers, they have no space inside them where the air can vibrate to make the sounds louder.

Drums are probably the oldest and simplest musical instruments in the world. They contain a space filled with air, and have a flexible drumskin stretched across one end. When the drumskin is struck, it vibrates and makes a noise.

Although all drums work in basically the same way, they can still make a huge range of different noises. Size is important. A big bass drum makes a much deeper sound than a small bongo drum. The drum's pitch is also affected by the drumskin. A tight drumskin makes a higher note than a slack one.

## MAKE it WORK!
Experiment with the sounds and tones of different drums. Try out some of these ideas and, if you like, put together a collection of drums to make your own drum kit.

## You will need

| | |
|---|---|
| string | thick paper |
| balloons | rubber bands |
| glue or sticky tape | thin wooden dowels |

boxes and tins of all shapes and sizes
a sheet of plastic or an old plastic bag
an eyelet punch or cardboard hole reinforcers

### ▶ Bongo drums
**1** Cut cardboard tubes into several different lengths, to make bongo drums that will sound different notes.

**2** Cut some flat pieces of balloon rubber to make the drumskins, and fix across the tops of the cardboard tubes with elastic bands.

**3** Attach a dowel to each drum. That way you can hang them up, and the notes will sound more clearly from the open ends of the drums.

### ▲ Tin-can drums
**1** Take the lids off both ends of a tin can. Wash the tin, being very careful not to cut yourself on any sharp edges inside.

**2** Stretch balloon rubber over the ends of the can, and secure it with elastic bands.

## ▼ Biscuit-tin drums

Take the lid off a biscuit tin. Then cut a sheet of thick paper or plastic, slightly larger than the top of the tin. Cut slits around the edge of the paper or plastic, and glue or tape it down to make a drumskin across the top of the tin.

## ▼ Tuneable drums

**1** Take a large can or cardboard tube which is open at both ends.

**2** Cut two circular pieces of thin plastic sheeting. (You can buy this at a kite shop, or cut up an old bag made of thick plastic.) Make small holes around the edge of each circle.

**3** Now reinforce the holes in the plastic. If you can get hold of an eyelet punch, use that to press small metal rings in place around the holes. Otherwise, you can use reinforcing rings made out of sticky-backed card.

**4** Thread string through the holes, from one drumskin to the other as shown in the photograph on the right. When you pull the string tight, the pitch of the drum will be higher.

Whirling rattles like the one on this page used to be a common sight at sports events. Fans would wave their rattles to cheer on their team.

Our rattle makes a very loud noise. It creates sound in two different ways. The wooden surfaces of the cog and tongue make a clattering noise as they hit one another, while the whirling movement of the rattle sets off its own rhythmic pattern of sound vibrations in the air around it.

## You will need

four metal washers
a drill and hacksaw
thin wooden dowelling
thick wooden dowelling
a fairly thick piece of wood (about 5 mm or
   1/4 in thick) for the frame of the rattle
thinner wood for the tongue of the rattle
matchsticks or thin slivers of wood

a cotton reel
strong wood glue

**Be careful!** This project involves some quite difficult woodworking. Ask an adult to help you with all the stages when you have to cut the wood or drill holes.

## MAKE it WORK!

Make your own whirling rattle, take it out of doors and see how loud a noise you can make with it!

top and bottom of rattle frame

rattle tongue

thin dowel

pieces for end of frame

cotton reel for cog

slivers of wood          washers          thick dowel

**1** Cut two identical oblong pieces out of the thicker wood. These will form the top and bottom of the frame of the rattle.

**2** Cut an oblong of thin wood, slightly shorter than the width of the frame. This will be the tongue of the rattle.

**3** Cut the thick dowel in two – you need a long piece for the handle of the rattle, and a much shorter piece for the top.

**7** Assemble the rattle as shown in the photograph on the left. Thread the thin dowel through the cog and slip a washer on either side. Then add the top and bottom of the frame, followed by two more washers. Fit the ends of the thin dowel into the holes you have drilled in the thicker dowel.

**8** Now glue the tongue and the outer end of the rattle in place. You must position the tongue very carefully. It should just touch the cotton-reel cog, so that it makes a noise, but should still be springy enough to let the cog spin around freely.

**4** Cut two small rectangles of thick wood. They fit across the outer end of the rattle, with the tongue sandwiched between them.

**5** Make the cog wheel. Cut several thin slivers of wood, or cut the heads off several matchsticks. Then glue these sticks around the side of the cotton reel, using strong wood glue.

**6** Take a drill the same width as the thin wooden dowel. Drill a hole through one end of the top and bottom pieces of the frame. Then drill the same size holes into the centre of both pieces of thick wooden dowel.

Assembling the whirling rattle

**Be careful!** Make sure that you use very strong wood glue to stick the pieces of the rattle together. Follow the instructions on the packet carefully, and try not to let the glue touch your skin.

The washers in the whirling rattle stop friction between the pieces of wood and help the rattle swing more easily.

# 38 Bangs & Hums

Have you ever burst a balloon by accident? The loud, unexpected bang can make you almost jump out of your skin! The air trapped inside the balloon suddenly rushes out, setting off sound waves which we hear as a loud bang. The sound of any explosion is caused by air moving at great speed – that's why the sound blast from a bomb will knock over people and even buildings.

## MAKE it WORK!

Moving air will make all sorts of noises. It can hum, whistle and whirr, as well as bang. Put together these simple cardboard gadgets and see how they crash like a thunder clap or hum like a bumble bee.

### You will need

| | |
|---|---|
| coloured card | glue |
| brown wrapping paper | string |
| scissors or a craft knife | a metal washer |

### Making a banger

**1** Cut a square of card.

**2** Cut another square of brown paper, slightly larger. Snip this square in half from corner to corner to make a triangle.

**3** Place the brown paper on top of the card as shown, and fold it where it overlaps the card. Glue paper and card along the overlaps.

**4** Fold both card and paper from corner to corner.

**5** To make a sound, hold the corner of the banger as shown on the left. Sharply tug it downwards. The paper beak will flick out and make a bang.

### ▲ Making a whirrer
**1** Cut out the two cardboard shapes shown on the left, to make the wing and flapper sections. Slot them together.

**2** Pierce two holes in the wing as shown, and glue a small metal washer to the nose to weight it down. Thread about a metre of string through the two holes.

**3** Take the whirrer out of doors and spin it round your head. At first, the wing will just make a clattering noise, but as you speed up the sound will change to a high-pitched whirr.

### ▼ Making a spinner
**1** Cut two hexagons of coloured card. Make two small holes at the centre, and four or five larger holes around the edges.

**2** Glue the hexagons together. Thread a piece of string through the centre holes and tie the ends together.

**3** Twist the string round and round. Then pull it outwards. As you pull, the string will keep on winding and unwinding itself and the spinner will keep on humming.

**String instruments** make music when a string or cord is plucked, so that the air around the string starts to vibrate. The pattern of a string instrument's sound waves depends on three things: the length of the string, what a string is made of and how tight it is.

A vibrating string on its own does not make much noise, and to make the sound louder, most string instruments have a **soundboard** and **resonator**. The soundboard picks up the strings' vibrations and transfers them to the resonator – a big space filled with air, which amplifies the sound.

**You will need**
an old tea chest
a broom handle
a hammer
string
a drill
nails

## Make a tea-chest bass
Make your own instrument with one adjustable string and a tea-chest resonator.

**1** Ask an adult to help you drill a hole in one corner of the tea chest. The broom handle should fit loosely inside the hole.

Drilling the hole

Fitting the handle

**2** Tap nails into the opposite corner of the tea chest and the top of the broom handle. Tie a piece of string between the two nails.

**3** Put one foot on the chest to hold it steady and pluck the string. Pull the handle back to tighten the string and make higher notes. Push it forwards for lower notes.

## Make a stretched-string zither

Each string on this zither is stretched by the weight of a water bottle so that it makes its own individual note.

### You will need

a hacksaw
a metal strip
food colouring
eight bottles of the same size
eight equal lengths of string or fine twine

a long wooden board
a hammer and nails

**1** Tap a row of eight nails into the middle of the wooden board. Space them out equally across the board.

**2** Ask an adult to help you cut a strip of metal the same width as the board. Make eight equally-spaced grooves in the strip.

**3** Cut a groove across the wooden board, near one end, and wedge in the metal strip.

**4** Dye the pieces of string in eight different colours with food dye. Fasten each string to a nail and run it across the metal strip and over the end of the board.

**5** Tie the loose end of each string tightly around the neck of a bottle. Then pour a different amount of water into each bottle. The more water in a bottle, the tighter the string will be, and the higher its note. Try to make an octave scale.

**6** Add food colouring to the water, to match the different strings.

## Playing strings

Many string instruments call for nimble fingers. Violinists or guitar players, for instance, hold down the strings in different places along the necks of their instruments. By altering the length of the strings they make different notes.

## Make a banjo

This home-made banjo is almost like the real thing. It has a **bridge** to transfer the strings' vibrations to the soundboard, and a margarine tub which acts as a resonator. See how many different notes it will play.

## You will need

a plastic tub
a strip of metal
wood and string
a bradawl and hacksaw
a screwdriver and screws

**1** Cut a length of wood for the soundboard, about three times the width of the plastic tub.

**2** Ask an adult to cut four small grooves in the metal strip, to make a bridge. Then cut another groove across the soundboard, near the bottom, and fix the bridge in it.

**3** Mark holes in the wood where the screws will fit, following the photograph below. Twist the screws down a little way, and tie on the strings, leading them across the metal bridge.

**4** Ask an adult to cut another piece of wood, the same depth as the plastic tub, to support the soundboard. Glue this support and the soundboard in place.

**5** Twist the screws down further in order to tighten up the strings and fine tune the banjo.

## Matchbox guitar

Stretch four rubber bands round a matchbox as shown on the left, and insert a bridge of thick card. The angle on the bridge means that each band is stretched by a different amount, so that it sounds a different note.

## Making guitars

Rubber-band instruments are easy to make. Try either a large shoebox guitar, or the smaller, portable matchbox version.

## Shoebox guitar

To make the shoebox guitar, just cut a hole in the lid, and stretch the bands across. Make the bridge from a sandwich of thick card and coloured paper as shown below. It should be so strong that you can slide it along to make the strings tighter.

### You will need
a shoebox
matchboxes
a craft knife or scissors
rubber bands of all sizes
thick card and coloured paper

The shoebox guitar makes a deeper, richer sound than the little matchbox models, because the resonator is bigger and the strings (the rubber bands) are longer.

Nowadays, we can record music on cassettes or compact discs, but for hundreds of years, the only way of recording music was to write it down on paper. Over the centuries, a complicated system for writing music developed. This system of **musical notation** is now used for western music in many different countries.

The symbols and instructions on a **score** give all kinds of information to the musician – which notes to play and what rhythm to follow, how loud or soft, and how fast or slow the piece is. Learning to read and write music is like learning a new language.

▼ **Western** music is usually written on a set of five lines and four spaces called a **staff**. The positions of the notes on the staff tell the musicians which notes to play.

*Some of the instructions on a musical score – especially those about the speed of a piece – are usually written in Italian. Andante, for instance, means slowly. Prestissimo means as fast as you can.*

▲ To check that they are playing a piece at the right speed, musicians use a clockwork timer called a metronome. It has an upright pendulum that ticks backwards and forwards with a regular beat. By moving the weight up or down the pendulum, it can be made to tick more slowly or quickly.

## MAKE it WORK!

You can work out your own simple system of writing down music and use it to record any music you compose for your home-made instruments.

**1** Colour in the stick-on spots to match the colours of the stretched-string zither on page 41 or the bottle organ on page 26. The different coloured spots will stand for the different notes.

**2** The size of the spots tells you how loudly to play the instrument. Small spots mean you play the note softly *(piano)*, medium-sized spots mean medium loud *(mezzo forte)* and large spots are loud *(forte)*.

**You will need**
graph paper
white stick-on spots in three sizes

▶ Here's a simple example, going up the octave scale and down again, not too loud and not too soft.

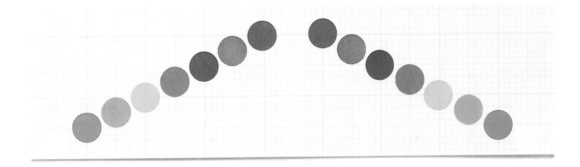

▶ Up and down the scale again. Softly on the way up, loudly on the way down.

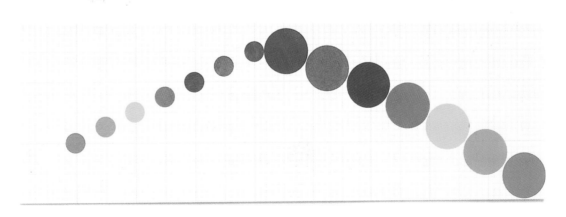

▶ Music for two players. The line stands for a drum beat every other note. The guitar plays orange and red notes, getting louder and louder.

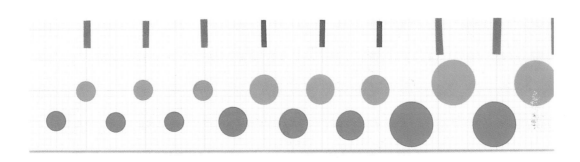

**Aerial** A long piece of wire or metal that picks up radio waves. All radios and televisions need an aerial.

**Amplify** When you amplify something, you make it louder.

**Bridge** In a string instrument, the bridge is a small, raised piece of wood or metal that connects the strings to the soundboard.

**Cochlea** A curly tube inside the ear, filled with liquid and lined with tiny hairs. It turns the vibrations made by sound waves against the ear-drum into nerve pulses.

**Decibel** Decibels are the unit for measuring how loud or soft a sound is.

**Diode** A tube-shaped device used in electronic circuits. The diode stops the current flowing both ways through the circuit and makes it flow in just one direction.

**Ear-drum** The part of the ear inside the head that vibrates when struck by sound waves.

**Hertz** Sound waves are measured in a unit called the Hertz. It is named after the German physicist, Heinrich Hertz.

**Incus** One of the three main bones inside the ear. Often also called the anvil bone.

**Malleus** One of the bones inside the ear. It is often also called the hammer bone.

**Musical notation** Recording music by writing it down on paper.

**Nerve pulses** Nerve pulses are signals sent out by our organs of sense, such as our eyes and ears. These signals pass along our nerves, to the brain, where they are decoded as sights, sounds, tastes, smells or feelings.

**Note** In music, a note is a sound played at a particular pitch.

**Octave** An octave is a scale of eight notes that rise and fall step by step. Most western music is based on this octave scale.

**Percussion instruments** Musical instruments that are played by being struck or shaken. Drums, maracas and xylophones are all examples of percussion instruments.

**Phonograph** An early kind of record player. It recorded sounds by making a series of bumps and dips in grooves on a rotating cylinder.

**Physics** The branch of science that finds out about different kinds of energy and matter. Sound is one of the forms of energy investigated by physicists.

**Pitch** The pitch of a note is how high or how low it is. If sound waves are close together they make a higher pitch than if they are far apart.

**Radio waves** Waves of electromagnetic energy that travel rapidly across long distances. Sounds are turned into radio waves by a radio transmitter, and changed back into sound again by a radio receiver.

**Resonator** A space inside an instrument filled with air, which vibrates and makes the sound of the instrument louder.

**Scale** A series or sequence of different musical notes.

**Score** A written record of a piece of music.

**Sound waves** Sounds travel through the air in waves. When someone bangs on a drum, for example, the drumskin squashes together the air beside it every time it vibrates. If we could see air, the movement of sound waves would look rather like the ripples on a pond when a stone is thrown into the water.

**Soundboard** The board in a piano or a string instrument which is connected to the strings by the bridge. It vibrates when the strings are struck or plucked.

**Staff** In music, a staff is a set of five lines on which the notes are written.

**Stapes** One of the bones of the middle ear. It is often also called the stirrup bone.

**Stethoscope** An instrument used by a doctor to listen to the inside of a patient's body.

**String instruments** String instruments are played by plucking strings or scraping them with a bow. A sitar, a zither and an electric guitar are all examples of string instruments.

**Vacuum** A completely empty space, that does not even contain air. Outer space is a vacuum.

**Vibrate** When something vibrates, it moves backwards and forwards a small amount very quickly. A guitar string vibrates if you pluck it.

**Wind instruments** Instruments that are played by blowing through them. Trumpets and recorders are both wind instruments.